Willian Paul Palacios Vargas

# Evaluation of Chitosan as a Heavy Metals Leachate Remover

**Willian Paul Palacios Vargas**

# Evaluation of Chitosan as a Heavy Metals Leachate Remover

## Landfill of Canton Mejia

**ScienciaScripts**

**Imprint**

Any brand names and product names mentioned in this book are subject to trademark, brand or patent protection and are trademarks or registered trademarks of their respective holders. The use of brand names, product names, common names, trade names, product descriptions etc. even without a particular marking in this work is in no way to be construed to mean that such names may be regarded as unrestricted in respect of trademark and brand protection legislation and could thus be used by anyone.

Cover image: www.ingimage.com

This book is a translation from the original published under ISBN 978-620-2-14947-1.

Publisher:
Sciencia Scripts
is a trademark of
Dodo Books Indian Ocean Ltd. and OmniScriptum S.R.L publishing group

120 High Road, East Finchley, London, N2 9ED, United Kingdom
Str. Armeneasca 28/1, office 1, Chisinau MD-2012, Republic of Moldova, Europe
Printed at: see last page
**ISBN: 978-620-6-34116-1**

Table of Contents :

**SEK INTERNATIONAL UNIVERSITY**


**FACULTY OF NATURAL AND ENVIRONMENTAL SCIENCES**


**"EVALUATION OF THE EFFECTIVENESS OF CHITOSAN AS A REMOVER OF HEAVY METALS IN THE LEACHATES FROM THE LANDFILL OF CANTON MEJÍA, PICHINCHA, ECUADOR".**


Performed by:
**WILLIAN PAUL PALACIOS VARGAS**

Project Director:
**MSc. Emma Ivonne Carrillo Paredes**

# DEDICATION

I dedicate this research work to my mother who has been both father and mother at the same time, and is the person who through hard work and effort has succeeded in life, being an example to follow. Thank you mom for always being by my side, supporting me through thick and thin, and for having instilled in me values and principles that have served me well throughout my life. To my sister Cristina, with whom I have spent most of my time and who has been the person who has made great sacrifices for my well-being.

# THANK YOU

I would like to thank Professor Ivonne Carrillo for her wise direction of the final graduation project. Her professionalism and dedication were decisive in the completion of this document.
To the SEK International University, for its efforts to train outstanding professionals.
To be submitted to: *Revista de Ingeniería e Investigación*
*To be Submitted: Journal Engenering & Investigation*

## EVALUATION OF THE EFFECTIVENESS OF CHITOSAN AS A REMOVER OF HEAVY METALS IN THE LEACHATES OF THE LANDFILL OF MEJÍA CANTON, PICHINCHA, ECUADOR

Effectiveness of Chitosan as a heavy metal remover from the leachate of the Sanitary Landfill of Canton Mejia, Pichincha, Ecuador


Paul Palacios[1] & Ivonne Carrillo[2]

[1]Universidad Internacional SEK, Faculty of Natural and Environmental Sciences, Quito, Ecuador. Email: palaciosvargas 22@hotmail.com

[2]Universidad Internacional SEK, Faculty of Natural and Environmental Sciences, Quito, Ecuador. Email: emma.carrillo@uisek.edu.ec

[2]Correspondence Author: Universidad Internacional Sek, Faculty of Natural and Environmental Sciences, Quito Ecuador. Campus Miguel de Cervantes Carcelén Quito Ecuador ivonne.camllo@uisek.edu.ec


Running Title: Chitosan for Leachate Treatment

# SUMMARY

Currently, the final disposal of municipal waste (MSW) is a problem that affects people's health and the environment. Some controlled sanitary landfills have the disadvantage of producing liquids with high levels of contamination as a result of biological decomposition, called leachates, so it is necessary to apply environmentally friendly decontaminants to control these effluents that are discharged into water sources. Thus, given the importance of this problem, the present research, which was conducted in the landfill pool of the Mejía Canton, Province of Pichincha, has the objective of evaluating the effectiveness of chitosan as a remover of the different heavy metals contained in the leachates. These liquids were previously treated by the FENTON advanced oxidation method, which is applied by combining an oxidizing agent, Hydrogen Peroxide ($H_2O_2$) with a catalyst of Ferrous Sulfate heptahydrate [Fe ($SO_4$)].$7H_2O$. On the other hand, Chitosan is a linear polysaccharide extracted and modified from the shrimp exoskeleton that has multiple uses due to the presence of hydroxyl (OH) and amino (-NH2) functional groups, including the removal of heavy metals in leachates. The hydroxyl groups of chitosan will act as a chelating agent on heavy metal ions forming chelates. For the Chelation process, the following parameters were taken into account: pH, Chitosan dosage and agitation time, and the following heavy metals were measured: Cadmium, Chromium, Iron, Lead and Zinc. Chelation with Chitosan applied to the Fenton-treated leachate at pH 3, one gram of Chitosan and 15 min of agitation was effective in removing the following metals; Cadmium, Chromium and Lead. The final percentages of heavy metal removal by chelating with chitosan were as follows; Cadmium 60 %, Chromium 50 %, Lead 9 %, **Key words:** Chelating Agent, Fenton, Leachate, Heavy Metals, Chelation, Chelates, Chitosan.

# ABSTRACT

Currently, the final disposal of urban waste (USW) is a problem that affects the health of people and the environment. Some controlled sanitary landfills have the disadvantage of producing liquids with high levels of contamination due to biological decomposition called leachates, so it is necessary to apply environmentally friendly decontaminants to control these effluents that are discharged to water sources. Thus, given the importance of this problem, the present investigation was carried out in the sanitary landfill of Cantón Mejía, Pichincha Province, with the objective of evaluating the effectiveness of Chitosan as a remover of the different heavy metals contained in the leachates. These liquids were previously treated by the FENTON advanced oxidation method, which is applied by combining an oxidizing agent, Hydrogen Peroxide (H2O2) with a ferrous sulfate catalyst heptahydrate [Fe (SO4)] .7H2O. On the other hand, Chitosan is a linear polysaccharide extracted and modified from the shrimp exoskeleton that has multiple uses due to the presence of the functional groups: hydroxyl (OH) and amino (-NH2), among them the removal of heavy metals in leachates. The hydroxyl groups of Chitosan will exert the function of Chelating Agent on the ions of the heavy metals forming Chelates. For the chelation process, the following parameters were taken into account: pH, chitosan dose and stirring time, and the following heavy metals were measured; Cadmium, Chrome, Iron, Lead and Zinc. Chelation with Chitosan applied in the leachate treated with Fenton at pH 3, one gram of Chitosan and 15 min of agitation was effective to remove the following metals; Cadmium, Chrome and Lead. The final percentages of the removal of heavy metals applying Chelation with Chitosan were the following; Cadmium 60%, Chrome 50%, Lead 9%.

**Keywords:** Chelating Agent, FENTON, Leachate, Heavy Metals, Chelation, Chelates, Chitosan.

# INTRODUCTION

In the last decade, Ecuador has suffered an increase in waste generation as a result of population growth, and poor waste management has led to the production of high amounts of contaminating liquids called leachates that are not properly treated.

Today's society and its technological development have increased the production of both urban and industrial wastes, the latter are characterized by, for the most part, containing substances that attribute to them hazardous and toxic characteristics, which require further and detailed analysis (Coral, 2011). Toxic and hazardous wastes are generated by human actions and are discarded because they are categorized as unusable, even though they contain in their structure materials that are highly hazardous to human health and the environment. Although today's society, and especially industrial activity, generates waste, it is essential that it be managed in the most environmentally acceptable way (Coral, 2011).

Heavy metals are one of the main pollutants that affect the health of the population and the environment, these are commonly treated with inorganic chemicals, hence the importance of research in applying a polysaccharide with biodegradable properties called Chitosan. The present study took as reference the "Romerillos" landfill located in the Mejía Canton, Pichincha Province, for which it was necessary to evaluate the effectiveness of Chitosan as a remover of heavy metals in the leachates of the landfill of that canton.

In most countries of the world there is a great concern to achieve what is called the integral development of human beings and all their activities, this process being possible only if the needs of society and the environment are addressed.

A major environmental problem at the Mejía Canton landfill is that the leachate collected in the pool is pumped into collection buckets and permanently recirculated. Another problem that plagues the landfill is the heavy metals present in the leachate produced by the type of waste that is deposited there. Heavy metals have been identified as one of the factors affecting the health of the population and the environment due to the toxicity of certain hazardous chemicals that are commonly used in the home. Table 9 of Ministerial Agreement 097, TULSMA, Book VI, Environmental Quality Standard and Effluent Discharge to Water Resources, stipulates quality parameters that effluents must meet prior to discharge into a freshwater body.

The generation of toxic and hazardous waste that occurs every day among the population has caused an increase in the concentration of heavy metals in leachates and the use of chemical products to retain heavy metals in contaminated water has generated a disadvantage, since they are not biodegradable and leave residues without any treatment, hence the importance of the use of chitosan to, in a certain way, maintain control and prevent their toxicity from increasing.

The heavy metals in the leachates from the Cantón Mejía landfill are found in concentrations that exceed the permissible limits established by Ecuadorian environmental regulations, and therefore represent a problem that must be studied, treated and solved by applying measures to reduce them.

In order to propose a solution to this problem, the following project was carried out at the Mejía Canton Landfill and at the SEK University laboratory using leachate samples collected at the landfill. New trends in infrastructure and respect for the environment point to the incorporation of Chitosan studies, with the objective of identifying, analyzing and evaluating its effectiveness in the removal of heavy metals in leachates.

According to the above mentioned, the present project sought to validate the application of chitosan as a chelating agent in the leachates of the Mejia landfill pool, as a methodology to reduce the contractions of different types of heavy metals, and the results obtained will provide new elements of judgment for authorities such as the GAD, to take new actions in the prevention and control of pollution, as well as in the application of environmental legislation.

The analyses performed by UISEK researchers and the support of the equipment provided by the same institution, allowed the research to be feasible and economical, also, the duration time of the experimental work was short term, and the last existing changes and/or modifications in the test trials performed prior to the present project were characterized *General Objective.*

To evaluate the effectiveness of the application of chitosan through chelation tests for the removal of heavy metals in the leachates of the Mejia landfill, Province of Pichincha.

*Specific objectives*

To determine the optimal dose of chitosan in the treated leachate from the Mejia landfill pool using chelation methods for the removal of heavy metals.

To reduce the concentration of organic matter (COD) present in the leachate from the Mejia County landfill pool, through the application of the FENTON Advanced Oxidation physicochemical treatment.

To evaluate the quality of chitosan obtained from shrimp exoskeleton through chelation test results.

Evaluate the effectiveness of Chitosan in treated pool leachate by Chelation test assays to remove heavy metals.

*Justification*

Different leachate treatment studies carried out at the Mejía landfill have concluded that they reduce the amount of organic matter, but the concentration of heavy metals has remained intact, with the main source being the different types of waste containing toxic and hazardous chemical compounds that are disposed of in the landfill.

Due to the fact that the required importance has not been given, heavy metals have become one of the greatest environmental problems, as well as the different physical health diseases caused by the toxicity of each of their components, which have caused an alteration in the normal behavior of human beings.

Among the general affectations that have a major impact on human beings are: lung cancer, necrosis of the genital organs, vomiting and nausea, arterial hypertension, saturnism, mental retardation, deafness, central and peripheral nervous system affectation, kidney affectation, constipation, loss of appetite, anemia, paralysis, headaches, neurosis, schizophrenia and asbestosis (Coral, 2011).

All these facts influence the exposure of these toxic pollutants, for this reason it is necessary to carry out more detailed and accurate studies and evaluations in order to adopt them, as well as to implement corrective measures. Hence the importance of evaluating the effectiveness of Chitosan as a biodegradable chelating agent capable of removing different types of heavy metals carried out by the UISEK research group.

This research tested, compared, verified and evaluated the effectiveness of treatment with chitosan on the different types of heavy metals contained in the leachate from the Mejia landfill.

*Background*

Canton Mejía is located in the southeast of the province of Pichincha, with a surface area of 1,459 $km^2$ . The urban area is between 20% and 30% consolidated, with a population density between 50 and 60 inhabitants/$km^2$ . Its altitude with respect to sea level is between 600 and 4,750 m.a.s.l. (Barros, D. and Ortiz, J. 2010).

The Mejía Canton landfill is located 13 km south of Machachi, with an area of six hectares of land, of which only 10% is currently occupied. The amount of solid waste entering the landfill's Recycling and Composting Center is 46 tons/day, according to information provided by the Municipality of Mejia (Zaldumbide, 2013).

The last landfill in operation has a leachate collection pool that is 17 meters long, 9 meters wide and 4.5 meters deep. At the base of the pool there is a geo-membrane that separates the liquid from the soil. This is located in the lower part of the landfill, so the liquids are deposited by gravity.

Currently, the treatment plant is not operational, so the leachate collected is pumped to collection tanks and recirculated indefinitely. This makes it necessary to design an efficient treatment so that the concentration of the components does not affect the environment. The designer of the leachate treatment plant estimated the daily generation of these liquids according to the surface area of the ponds. The average leachate generation values are shown in **Table 1.**

**Table 1.** Generation of Leachate in the "Romerillos" Landfill, Canton Mejia.

| Canton | Surface area (m )$^2$ | Average (m$^3$ /day) |
|---|---|---|
|  | 15000 | 7,12 |
| **Mejia** | 30000 | 14,52 |

| | 45000 | 21,64 |
| | 60000 | 29,04 |

**Prepared by:** Paul Palacios, 2018. **Source:** Marcelo Castillo P. 2003.

The design of the leachate treatment plant has a capacity of 29.04 m$^3$ /day. The constructed (non-operational) treatment plant has two separate structures. The first is located at the entrance to the leachate pool and has a liquid pumping system and aeration trays (Zaldumbide, 2013).

The second structure is located 100 meters from the leachate pool and, like the first, has a pumping system, six aeration trays, a sedimentation tank and an equalizer (Zaldumbide, 2013).

*Theoretical Framework*

The progress in the research carried out for the removal of heavy metals using chitosan has been limited, due to the fact that in many studies it has not been possible to carry out tests for all the metals contained in the leachates. Therefore, it has not been possible to adopt chelation methods that can be used as a useful tool for large-scale treatment.

Four studies show that Chitosan in liquid, gel and solid state are effective in different states in the removal of the following metals; Cadmium, Copper, Chromium, Mercury and Lead.

The first study carried out at the Central University of Ecuador, shows the methodology and results of the application of chitosan-glutaraldehyde (QGD) and chitosan-ferric nitrate (QFe) for the removal of lead in contaminated water. In the chelation stage, pH, adsorbent dosage, agitation speed, temperature and contact time were taken into account as factors in the adsorption of lead. The result was 98.40% efficiency (Dávila & Bonilla, 2011).

The second study carried out at the University of Guadalajara, shows the application of Chitosan in the form of alginate-chitosan hydrogel beads and alginate-chitosan sulfate, for the removal of metallic copper ions. The methodology applied was to measure the size of the beads, observe the morphology and the removal capacity at two pH values, With this technique it was demonstrated that Chitosan in the state of alginate hydrogels can remove more than 80 % of copper ions and in the form of alginate-Chitosan Sulfate, removed all copper (Balleño, Ríos, Aranda, Morales, Mendizábal & Katime, 2016).

The third study carried out at the Instituto Tecnológico de Costa Rica was based on the application of chitosan in the form of membranes in industrial wastewater. For this study, physical-mechanical tests were carried out on membrane thickness, pressure resistance, and membrane tension. The results were obtained using atomic absorption equipment for cadmium, copper and chromium, giving the following removal results: 23.05% for chromium, 14.10% for cadmium and 16.09% for copper. Thus, the effectiveness of the membranes in the removal of heavy metals from wastewater did not exceed 50%, so it is recommended to use a natural substance or biopolymers (Villalobos, 2011).

Finally, the fourth study carried out at the Universidad Veracruzana was based on the application of chitosan from shrimp exoskeletons with a deacetylation degree of 75% in the removal of lead ions from aqueous solutions. Adsorptions were performed at concentrations of 5, 25, 50, 50, 70, 100 and 120 mg/L and the removal was 4, 20, 45, 60, 80 and 90 mg/L respectively at temperatures of 25, 35 and 50 °C with an amount of Chitosan of 0.1 g for each solution. The adsorption effectiveness was more efficient at 25 °C, with an adsorption capacity of 107.41 mg $Pb^{2+}$ per gram of Chitosan. The adsorption efficiency percentages were shown to be above 80%, demonstrating that the process is suitable for the removal of metals in water using Chitosan (Altamirano, 2015).

Currently, no new methodologies have been applied to determine the effectiveness of chitosan for the removal of heavy metals in leachates from landfills in Ecuador, for this reason, the group of researchers from SEK International University has proposed to develop and establish a methodology to remove heavy metals in leachates, taking into account different parameters and performing chelation tests to establish it as a methodological basis.

The main parameters taken into account for the development of the project were as follows:

- pH

- COD

- Cadmium

- Chrome

- Iron

- Lead

- Zinc

To validate the data of the concentration of heavy metals obtained experimentally using the flame atomic absorption spectrophotometer, the following parameters were taken into account; Volume, pH, dose of Chitosan powder, agitation time and agitation speed. With these parameters it is intended to establish a method of treatment of heavy metals in leachates, specifically for those of the Mejia Canton landfill.

Many of the contaminants have weak electrolytes, so the adsorption process depends mainly on the pH of the solution. It is recommended that the solution has an acid pH between 2 and 3, since the amine groups of chitosan are more easily pronounced, which causes the electrostatic repulsion of metal ions (Díaz, 2012).

This leads to competition for adsorption sites between protons and metal ions. This is why pH can affect the speciation of metal ions, and the change in metal speciation can result in converting the Chelation mechanism into an electrostatic attraction mechanism (Walco, 1997).

*Concepts*

**Shrimp Production**

At present, the shrimp industry in Ecuador ranges between 180,000 hectares along the coasts of the country. Shrimp farms are distributed in the provinces of: Esmeraldas, Manabí, Guayas and El Oro (Loaiza, 2016).

According to statistics collected from the National Chamber of Aquaculture, exports in the last two years are around 50 thousand pounds per month (Acuacultura, 2015).

Much of the waste, such as the head and carapace, is disposed of in the river and in the garbage, generating bad odors and allowing it to become toxic waste for health and the environment.

**Chemical Oxygen Demand**

Known as COD, it measures the amount of organic and inorganic matter present in liquids that are susceptible to chemical oxidation. This parameter is fundamental for water control and treatment, since it is an indicator of environmental contamination.

For COD it is necessary to use a chemical reagent with a high oxidizing power such as potassium dichromate. COD is measured in mg $O_2$/L. Prior to COD measurement, it is essential to use a digester at a temperature of 150 °C (Jiménez, 2012).

**Leachate**

These are liquids formed as a result of the decomposition of solid urban waste deposited in sanitary landfills and the percolation of rainwater that falls on them. The characteristics of leachates depend on climate, temperature, moisture content, age of the landfill, rainfall regime, type of cover and density of the landfill mass (Morales, 2007).

The composition of a leachate is characterized by high amounts of organic matter (biodegradable, but also refractory to biodegradation), organic and inorganic salts, nitrogen, heavy metals and other diluted chemicals, varying with the age of the landfill (Steiner, 2008), the characteristics of the waste deposited, the meteorology of the site and mode of operation (Castrillón, 2008), (Renou, 2008).

Leachates contain all the characteristics to be considered a major pollutant, i.e., high content of organic matter, nitrogen and phosphorus, abundant presence of pathogens and also toxic substances such as heavy metals and organic constituents (Giraldo, 2001). The organic compounds present in leachates are: proteins, carbohydrates, alcohols, and mainly volatile fatty acids; additionally, leachates contain a large amount of ammoniacal nitrogen (Torres, 2005).

These volatile fatty acids are easily diluted in the landfill leachate, lower its pH and contribute to the solubilization of metals present in the waste disposed in the landfill (Morales, 2007).

The concentrations of all parameters in the young leachate are much higher than in the old leachate. The BOD/COD ratio for a young leachate is high indicating good biodegradability, while for an old leachate it is low indicating low biodegradability of organic matter. The concentrations of dissolved salts and heavy metals are much higher in a young leachate, generating toxicity problems when using biological processes for BOD removal (Renou, 2008).

**FENTON Advanced Oxidation Method**

This method has been proposed as an alternative in recent years for soil and water decontamination with amazing results. It has a high efficiency and low cost to remediate waters contaminated with toxic compounds. The FENTON method can degrade aliphatic compounds, chlorinated aromatics, PCBs, nitroaromatics, dyes, chlorobenzene and phenols, with the exception of acetone, or acetic acid. The FENTON method has been effective in reducing COD in municipal water, groundwater, leachate from municipal landfills and paper and textile companies. This system works by combining hydrogen peroxide ($H_2O_2$) and iron in acidic solutions, and the dosage depends on the COD value of the liquid to be treated.

This process consists of the addition of iron salts in the presence of $H_2O_2$, in an acid medium, for the formation of °OH radicals. In addition to the formation of °OH radicals, perhydroxyl radicals ($HO_2°$) are generated, which initiate an oxidation chain reaction to eliminate oxidizable matter. However, $HO_2°$ radicals are less oxidizing than °OH radicals (Domenech *et al.*, 2004).

<u>Heavy metals</u>

These are those metallic chemical elements that are toxic and poisonous at low concentrations, have high density, high molecular weight and are dangerous because they bioaccumulate in the organism. They do not degrade in natural environments, so chemical and costly methods are applied for their elimination (Cañizares, 2000).

**Cadmium**

Cadmium is a metal used in the electroplating industry, resistant to corrosion and is used for electroplating with steel and iron. The routes of cadmium absorption are inhalation and ingestion. After cadmium absorption via ingestion, it is transported to the liver, where it initiates the production of a low molecular weight protein called metallothionein. The symptoms produced by ingestion of cadmium in concentrations above 15 mg Cd/L are: nausea, vomiting, abdominal pain and diarrhea (Nordberg, 2007).

**Chrome**

Chromium is a metal used in a wide variety of equipment, including the manufacture of vehicle parts and electrical equipment. Valence VI chromium has acidic and oxidizing properties and is used as a compound in commercial and industrial applications. Chromium is capable of forming colored and insoluble salts. The main chromium compounds with valence VI are sodium dichromate, potassium dichromate and chromium trioxide (Nordberg, 2007).

Ingestion or inhalation of chromium compounds are rapidly absorbed by the body. Chromium generates irritant and corrosive effects immediately after absorption through the mucosa. The main disease produced by chromium is renal necrosis (Nordberg, 2007).

**Iron**

Iron is a metal that has multiple uses, one of them being the manufacture of parts, steel and alloys together with other metals. It is also used in the oil industry to increase the density of the fluids used in drilling wells. The main disease caused by iron is lung cancer (Nordberg, 2007).

**Lead**

Lead is a heavy metal resistant to sulfuric and hydrochloric acid, but in nitric acid it dissolves slowly due to the presence of nitrogenous bases. The main sources of lead contamination are industries, oil and mining. Among the main pollutants are batteries, electronic devices and paints, which in high concentrations of lead are harmful to human health and the environment. Among the negative impacts on the environment is that they can remain for long periods and bioaccumulate in organisms (Nordberg, 2007).

Lead is classified as a persistent environmental toxicant; once absorbed, it passes into the bloodstream, where more than 95% binds to erythrocytes, causing increased fragility and reduced cell lifespan. Consequently, it can damage the brain, kidney, liver and reproductive system. Moreover, lead exposure has been correlated with sterility, miscarriages, and neonatal deaths (Morales, 2007).

**Zinc**

Zinc is used for the protection of steel by galvanization, pigment for the manufacture of paints, component in the manufacture of cosmetics and in the pharmaceutical industry. The ingestion of zinc solubilized in acid liquids and with the presence of iron, generates the following symptoms; fever, nausea, vomiting, stomach pain and diarrhea. Zinc salts that enter the body through the skin or by ingestion can cause intoxication. Zinc chloride causes skin ulcers (Nordberg, 2007).

**Leachate Pool**

The purpose of the leachate pool is to keep the leachate from the landfill impermeable. This infrastructure allows these liquids not to come into contact with the soil, because they can seep into the groundwater and contaminate the soil if they are not properly controlled (Árevalo, 2018).

**Chitosan**

Chitosan is a polysaccharide found in the exoskeleton (shell) of many crustaceans, including shrimp, and is obtained by hydrolysis of chitin in an alkaline medium, usually sodium or potassium hydroxide. This polymer has the advantage of being renewable, does not pollute the environment and does not affect the organism that uses it (Lárez, 2006).

Chemically, Chitosan is a copolymer of N-acetylglucosamine (acetylated unit) and glucosamine (deacetylated unit), these units are 2-acetamido-2-deoxy-D-glucopyranose ($C_8H_{13}O_5N$, 203.1925 g/mol) and 2-amino-2-deoxy-D-glucopyranose ($C_6H_{11}O_4N$, 161.1558 g/mol); joined by a glycosidic bond of the type $\beta\text{-}(1\rightarrow4)$; both chitin and Chitosan are formed by linear chains of glucopyranose monomers, the difference lies in the C-2, where chitin possesses an acetamido group (- $COCH_3$), while in Chitosan that group is deacetylated to leave free the amino group (-$NH_2$) (Altamirano, 2015).

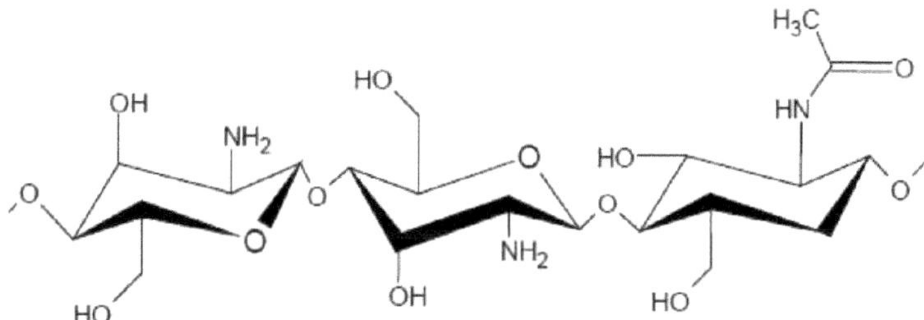



**Figure 1.-** Molecular **structure** of Chitosan

**Source:** Altamirano, 2015

Chitosan is an excellent absorbent due to its high hydrophilicity, chemical reactivity of the functional groups (-$COCH_3$, -$NH_2$ and OH), and the flexible structure of the polymeric chain.

The reactive groups of Chitosan, in particular, the -NH2 group selectively binds to virtually all group III transition metal ions, but does not bind to alkali and alkaline earth metal ions of groups I and II (Altamirano, 2015).

Chitosan has a cationic behavior, so the -NH2 group can be protonated by ion exchange. The presence of -NH2 groups in the structure of the Chitosan molecule makes this polymer a natural cationic polyelectrolyte with a pKa (strength to dissociate) of 6.2 to 7.0, depending on the degree of deacetylation of the polymer (Altamirano, 2015).

The fundamental properties of chitosan are solubility, viscosity, particle size, molecular weight, and degree of deacetylation. The capacity of chitosan to chelate metal ions depends on the degree of deacetylation, since this will determine the fraction of -NH2 groups that will be available to interact with the metal ions (Altamirano, 2015).

**Chelating Agent**

A chelating agent is a substance that forms strong complexes with metal ions forming non-covalent bonds. These can be inorganic and organic, and among the most commonly used are EDTA, citric acid, di- and triammonium citrate salts (Walco, 1997).

Only metals with a valence equal to or greater than +2 form chelates in the presence of ligands. Metal ions with valence +1 do not form chelates but salts with the ligand as an anion, i.e. a monodentate complex without a ring structure (Walco, 1997).

Chelating agents are often used to prevent one or more of the ordinary reactions of a metal ion without actually removing it from solution. For example, often a metal ion that interferes with a chemical analysis can be converted to a complex and thus eliminate its interference. In a sense, the chelating agent hides the metal ion. For this reason, scientists sometimes refer to these ligands as sequestering agents (Walco, 1997).

**Chelates**

Chelate means "clamp", i.e. the ring formed between the chelator and the metal is similar to the arms of a crab, the metal ion is trapped with two clamps (Walco, 1997). Single toothed ligands are those with only one atom including NH3 and Cl-. These can occupy a single site of a coordination sphere. Certain ligands have two or more donor atoms that can coordinate simultaneously to a metal ion, thus occupying two or more coordination sites. Polydentate ligands ("many-toothed" ligands). Because they appear to hold the metal between two or more donor atoms, polydentate ligands are also known as chelating agents (from the Greek word chele, "claw"). One such ligand is ethylenediamine (Walco, 1997).

## Chelation

It is defined as the formation of soluble complexes of metal ions in the presence of chemical agents that would normally produce precipitates in aqueous solutions (Labnews, 1997). Chelation can also be defined as the ability of a chemical compound to form a ring with a metal ion, giving a compound with different characteristics from the original metal. In this way the metal stops its normal chemical reactions (Walco, 1997).

# Chapter 1

## MATERIALS AND METHODS

Samples were collected in the leachate pool of the "Romerillos" landfill in Cantón Mejía, Pichincha province, where the following parameters were measured in situ: pH, conductivity, dissolved oxygen and temperature. **See Photo 1.**

**Photo 1.-** Landfill leachate pool
**Source**: William Palacios, 2018.

The development of this project was carried out in four methodological phases described below:

*Phase I. Chitosan Production*

Chitosan powder is produced using chemical methods from shrimp (*Penaeus vannamei)* shells supplied as raw material by seafood restaurants. Chitosan preparation steps are described below:

**Treatment of exoskeletons**

The exoskeletons were washed tenaciously with abundant water until the organic remains of the shrimp were eliminated, then dried at 70 °C in an oven for 8 hours, after which they were crushed, ground and sieved. The process began with 200 g of shrimp exoskeleton powder. The particle size obtained was 75 gm, since small particles increase the available surface area, which allows for high metal adsorption capacity and removal efficiency.

**Deproteinization**

Deproteinization was performed using 5% NaOH with constant agitation at 200 rpm and 100 °C for three hours considering a 1:10 solid-liquid ratio (w/v), the product obtained was washed with abundant distilled water until equal to pH 7. Finally, the exoskeleton was dried in the oven at 70 °C for 7 hours.

**Demineralization**

Demineralization was performed using 2 N HCl with constant agitation at 200 rpm and room temperature for three hours, in a 1:10 solid-liquid ratio (w/v), the product obtained was washed with abundant distilled water until equal to pH 7. Finally, the exoskeleton was dried in the oven at 70 °C for 7 hours.

**Purification**

Purification was carried out using NaOH 3% with constant stirring at 200 rpm and at a temperature of 100 °C for two hours, in a 1:10 solid-liquid ratio (w/v), the product obtained was washed with plenty of distilled water to equal pH 7 and was dried in the oven at 70 °C for 7 hours.

**Deacetylation**

At this stage the chitin obtained was treated using 50% NaOH with constant agitation at 200 rpm and a temperature of 100°C for two hours, in a 1:10 solid-liquid ratio (w/v), the final product obtained (Chitosan) was washed with abundant distilled water until equalized to pH 7. Finally, the Chitosan was dried in the oven at 70°C for 7 hours.

*Phase II. Leachate treatment with FENTON*

The leachate from the pool was treated with the FENTON Advanced Oxidation method to remove organic matter and its application depends directly on the COD value present in the leachate. From the leachate sample collected in the Mejía Canton landfill pool, 700 mL were extracted in a beaker and the following parameters were measured: pH and COD. The pH obtained was 8.5 and the COD value was 9790 (mgO2/L). Subsequently, the sample was adjusted to pH 3 with Sulfuric Acid and 33.48 mL H2O2 and 6.99 g of Ferrous Sulfate were added, these being the optimum doses for a ratio of 1:19. This process was carried out in a multiple agitation equipment for one hour at 100 rpm, then it was filtered and neutralized at pH 7 with 5% Sodium Hydroxide. Finally, a last filtration was carried out to obtain the final product and it was adjusted to pH 3 with Nitric Acid to be used in the Chelation tests. The following are the formulas for obtaining the optimum doses of H2O2 and FeSO4.

$$H_2O_2 \; = \; \frac{R1}{(1000 * Densidad)} \qquad \textbf{(Ec.2)}$$

$$H_2O_2 \; = \; \frac{14562,525 \text{ mg O2}}{(1000 * 1,45 \text{ g/cm}^3)}$$

$$R2 \quad = \quad 10,04 \text{ mL}$$

$$H_2O_2 \; = \; \frac{(R2 * 100)}{\text{Pureza } H_2O_2} \qquad \textbf{(Ec.3)}$$

$$H_2O_2 \; = \; \frac{(10,04 \text{ mL} * 100)}{30}$$

$$\mathbf{H_2O_2 = \quad 33,48 \text{ mL}}$$

**Formula 1.** Determination of the Optimal Dose of Hydrogen Peroxide (Anrango, 2018).

**Where:**

**Constant (k)** = 2.12

Prepared by: William Palacios, 2018.

**Formula 2.** Determination of the Optimal Dose of Ferrous Sulfate (Anrango, 2018).

**Where:**

**Molecular weight** FeSO4. **7** H2O = 277, 55 g/mol
**Molecular weight H2O2** = 34 g/mol
**Ratio 1** = 1
**Ratio 2** = 17

Prepared by: William Palacios, 2018.

*Phase III. Chelation*

$$H_2O_2 = k \cdot DQO \cdot Volumen \qquad \textbf{(Ec.1)}$$

$$H_2O_2 = \underline{2,12} * 9790 \text{ mg/L} * 0,7 \text{ L}$$

$$R1 = 14562,6 \text{ mg } O2$$

$$Fe_2 SO_4 = \frac{(R1 * \text{Peso Molecular FeSO}_4. \ 7 \ H_2O * \text{Relación 1})}{(\text{Peso Molecular } H_2O_2 * \text{Relación 2})} \qquad \textbf{(Ec.4)}$$

$$Fe \ SO_4 = \frac{(14, 562525 \text{ g } O2 * 277, 55 \text{ g/mol} * 1)}{(34 \text{ g/mol} * 17)}$$

$$\textbf{Fe SO}_4 = \textbf{6, 99 g}$$

The raw leachate sample before treatment is shown in **Photo 2**, and the FENTON-treated leachate sample is shown in **Photo 3.** Chelation tests were performed on the FENTON-treated leachate which has a COD of 1015 mg/l.

Photo 2.- Raw leachate from landfill pool
**Source**: William Palacios, 2018.

The conditions and values for realizing the effectiveness of chitosan in the removal of heavy metals in leachate are listed in **Table 2:**

**Table 2.-** Chelation Parameters

| No | Conditions | Value |
|---|---|---|
| 1 | Sample volume (mL) | 100 |
| 2 | Chitosan dosage (g) | 1 |
| 3 | Sample pH | 3 |
| 4 | Contact time (min) | 15 |
| 5 | Agitation speed (rpm) | 100 |
| 6 | Temperature ($^{o}$ C) | 20 |

**Source**: William Palacios, 2018.

Photo 3.- Leachate Treated with FENTON

**Prepared by**: William Palacios, 2018.

The metals to be analyzed in the FENTON-treated leachate are shown in **Table 3:**

**Table 3.-** Maximum Permissible Values for Heavy Metals

| No | Heavy Metals | Table 9, Ministerial Agreement 097 A, TULSMA, Book VI (mg/L) |
|----|-------------|------------------------------------------------------------|
| 1 | Cadmium (Cd) | 0,02 |
| 2 | Chromium (Cr) | 0,5 |
| 3 | Iron (Fe) | 10,0 |
| 4 | Lead (Pb) | 0,2 |
| 5 | Zinc (Zn) | 5,0 |

**Prepared by**: William Palacios, 2018.

Initially, 100 mL of FENTON-treated leachate sample was prepared at pH 3, then one gram of 75 |im Chitosan powder was added at room temperature, and kept in agitation at 100 rpm in multiple shaker equipment for 15 min.

The sample was filtered to separate the chitosan from the solution using a Buchner funnel adapted to a vacuum pump. This completed the chelation process.

Subsequently, the sample was digested with nitric acid to eliminate the organic matter impregnations in the metals, and finally, the sample was placed in the flame atomic absorption spectrophotometer and the following metals were measured: cadmium, chromium, iron, lead and zinc.

*Phase IV. Heavy Metals Measurement*

The main objective of digestion is the total decomposition of the sample to obtain a solution containing all the elements of interest. Inorganic substances must be completely transformed into soluble components and organic substances must be completely mineralized (Cahuasqui, 2011).

Prior to the measurement of heavy metals in the flame atomic absorption spectrometer, the samples were preserved and digested using the "Standard Methods" specified in **Table 4, 5 & 6.**

**Table 4.-** Samples to be digested

| No | Samples to digest |
|----|-------------------|
| 1 | Leachate treated with FENTON |

| 2 | |
|---|---|
| | FENTON Leachate treated with Chitosan |

**Prepared by**: william palacios, 2018.

**Table 5.-** Preservation of Samples using the Standard Method

| Sample Preservation |
|---|
| - Wash the sample beakers with soap and water, then wash them with distilled water and clean them carefully. |
| - The three samples should be stored in glass containers. |
| - Acidify the samples with HNO3 to pH=2. |
| - Refrigerate the acidified samples at 4 °C. |
| - Preservation of samples for heavy metal analysis lasts 6 months. |

**Prepared by**: William Palacios, 2018. **Source:** Standard Methods, 2005.

**Table 6.-** Preservation and Digestion of Samples by Standard Method

| Sample Digestion taken from SMWW, Issue 21, 2005 |
|---|
| - Prepare a volume of 100 mL for each sample in an Erlenmeyer flask and add 5 mL of HNO3. |
| - Place the three samples on the heating plate and evaporate them to 1/5 of the initial volume (20 mL). |

| | |
|---|---|
| - | Do not allow samples to dry out during digestion. |
| - | Allow the samples to cool to room temperature. |
| - | Add distilled water to the samples with residues and filter them. |
| - | Make up the solution with distilled water to 100 mL or initial volume. |
| - | Analyze the samples in the spectrophotometer. |

**Prepared by**: William Palacios, 2018. **Source:** Standard Methods, 2005.

# Chapter 2
# RESULTS

*Chitosan Production*

**Table 7.-** Obtaining Chitosan

| Obtaining Chitosan | |
| --- | --- |
| **Process** | **Quantity (g)** |
| Dried Cascara powder | 200 |
| Deproteinization | 123,8 |
| Demineralization | 72,9 |
| Purification | 58,8 |
| Deacetylation (Chitosan) | 40,6 |

**Prepared by:** William Palacios, 2018.

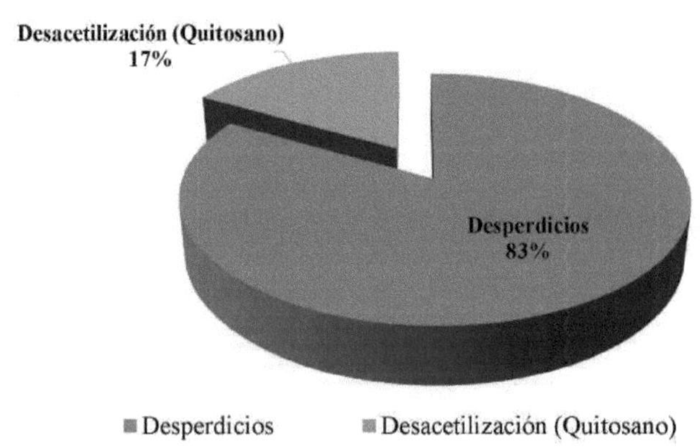


**Graph 1.-** Percentage of Chitosan Obtained and wastes

**Prepared by:** William Palacios, 2018.

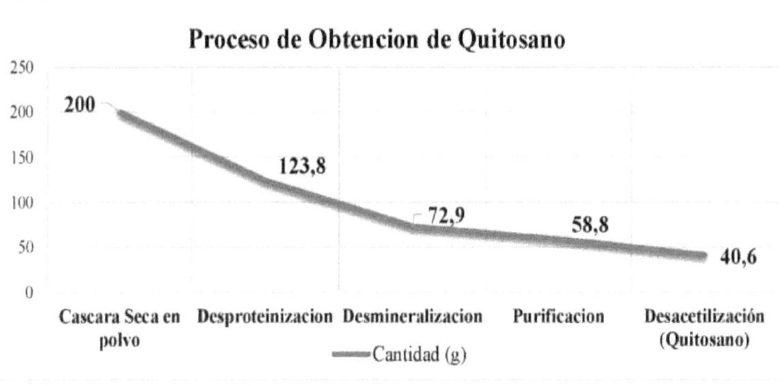



**Graph 2.-** Chitosan Obtaining Process

**Prepared by**: William Palacios, 2018.

To obtain chitosan, 200 g of shrimp shell were used, however, in each process there were losses as can be seen in Graph 2, deproteinization 123.8 g of product were obtained, demineralization 72.9 g, purification 58.8 g, and finally deacetylation 40.6 g of chitosan, i.e. only 17% of the sample is useful and 83% is loss.

*Fenton Method*

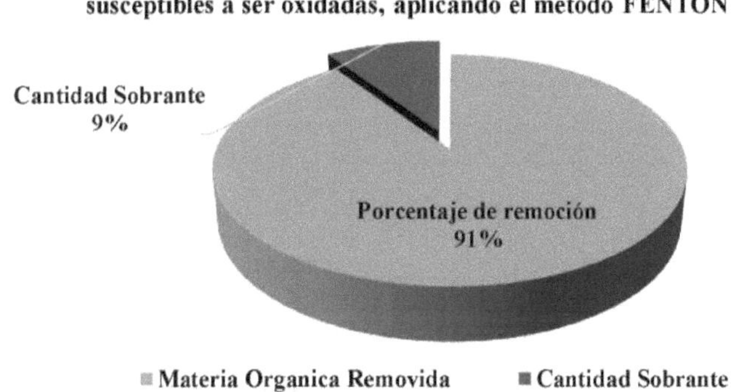



Percentage removal of organic matter and inorganic substances susceptible to oxidation using the FENTON susceptible to be oxidized, applying the FENTON method.

**Prepared by**: William Palacios, 2018.

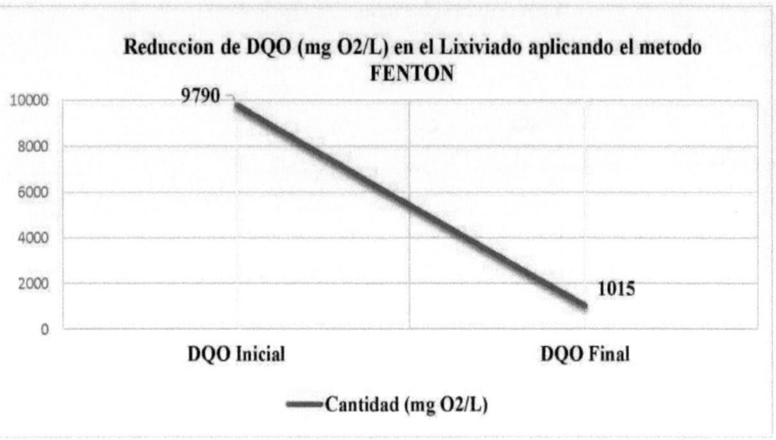


Reduction of COD (mg O2/L) in the leachate using the FENTON method.

**Prepared by**: William Palacios, 2018.

The leachate from the landfill pool has a COD of 9790 mg/L. The FENTON Advanced Oxidation treatment reduced the COD to 1015 mg/L, i.e. 91% of organic and inorganic matter was removed.

*Chelation*

**Table 8.-** Heavy Metals Concentration of each Sample

| Metal | Fenton sample (mg/L) | Sample Chitosan (mg/L) |
|---|---|---|
| Cadmium | 0,025 | 0,010 |
| Chrome | 0,030 | 0,015 |
| Iron | 17,16 | 17,16 |
| Lead | 0,174 | 0,159 |
| Zinc | 0,425 | 0,425 |

Prepared by: William Palacios, 2018.

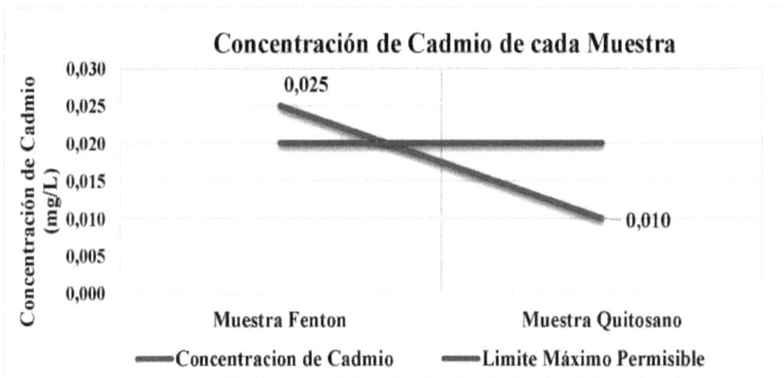



**Graph 5.-** Cadmium concentration of each sample.

**Prepared by**: William Palacios, 2018.

The leachate sample treated with FENTON had an initial concentration of 0.025 mg/L. Subsequently, applying the Chelation process the concentration of Cadmium was 0.010 mg/L. The initial sample exceeds the maximum permissible limit established for cadmium, which is 0.02 mg/L in Table 9 of Ministerial Agreement 097, TULSMA, Book VI, Environmental Quality Standard and Effluent Discharge to Water Resources, while the sample treated with chitosan is below this value.

**Graph 6.-** Chromium concentration of each sample.

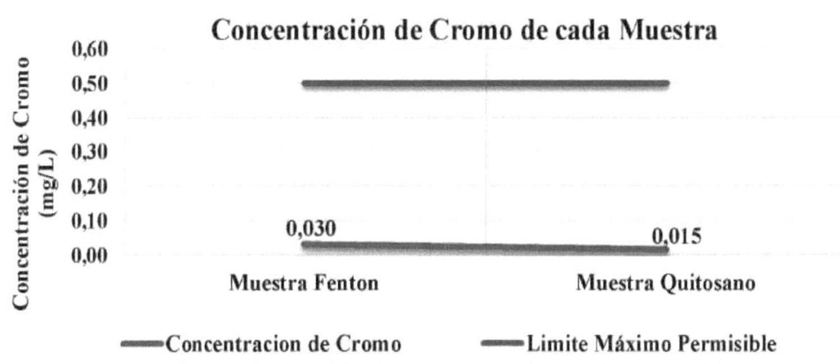



**Prepared by**: William Palacios, 2018.

The FENTON-treated leachate sample had an initial concentration of 0.030 mg/L. Subsequently, applying the chelation process, the concentration of Chromium was 0.015

mg/L. Both samples comply with the maximum allowable limit for chromium, which is 0.5 mg/L from Table 9 of Ministerial Agreement 097, TULSMA, Book VI, Environmental Quality Standard and Discharge of Effluents to Water Resources.

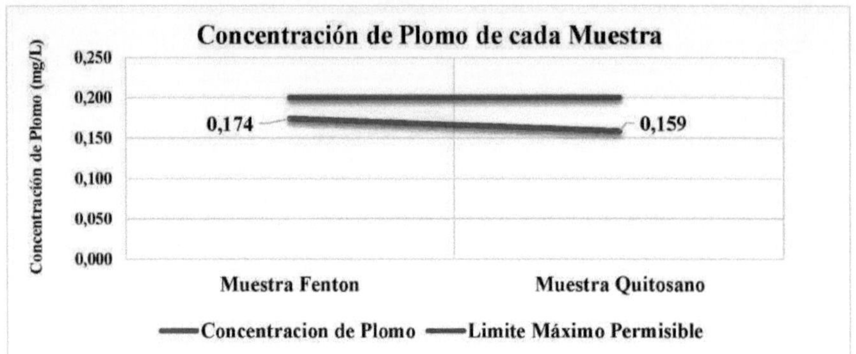


**Graph 7.-** Lead Concentration of each Sample

**Prepared by**: William Palacios, 2018.

The leachate sample treated with FENTON had an initial concentration of 0.174 mg/L. Subsequently, applying the Chelation process, the concentration of Lead was 0.159 mg/L. Both samples comply with the maximum permissible limit for lead, which is 0.2 mg/L, according to Table 9 of Ministerial Agreement 097, TULSMA, Book VI, Environmental Quality Standard and Discharge of Effluents to Water Resources.

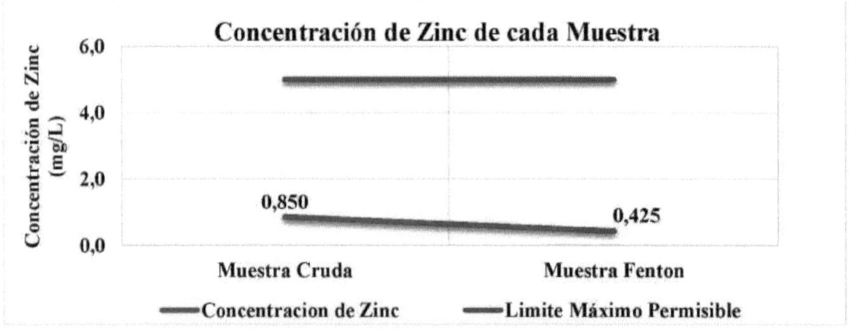


**Graph 8.-** Zinc concentration of each sample.

**Prepared by**: William Palacios, 2018.

The leachate sample treated with FENTON had an initial concentration of 0.425 mg/L.

Subsequently, applying the Chelation process, the Zinc concentration was the same 0.425 mg/L.

Both samples comply with the maximum permissible limit for zinc, which is 5.0 mg/L, according to

Table 9 of Ministerial Agreement 097, TULSMA, Book VI, Environmental Quality and Effluent

Discharge Standards for Water Resources.

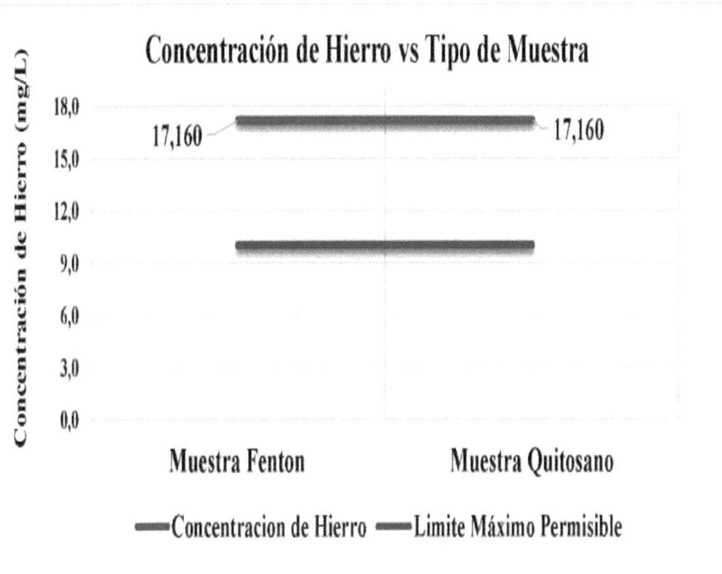


**Graph 9.-** Iron concentration of each sample.

**Prepared by**: William Palacios, 2018.

The leachate sample treated with FENTON had an initial concentration of 17.16 mg/L.

Subsequently, applying the Chelation process, the Iron concentration was the same 17.16 mg/L.

Both samples exceed the maximum allowable limit established for iron, which is 10.0 mg/L in

Table 9 of Ministerial Agreement 097, TULSMA, Book VI, Environmental Quality Standard and

Discharge of Effluents to Water Resources.

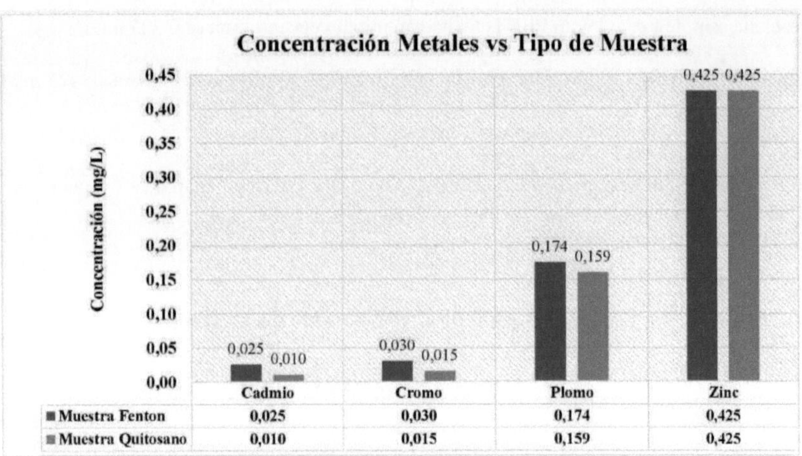


| | Cadmio | Cromo | Plomo | Zinc |
|---|---|---|---|---|
| Muestra Fenton | 0,025 | 0,030 | 0,174 | 0,425 |
| Muestra Quitosano | 0,010 | 0,015 | 0,159 | 0,425 |

**Graph 10.-** Ratio of Metal Concentrations of each Sample

**Prepared by**: William Palacios, 2018.

The lowest value of the leachate sample treated with Fenton is 0.025 mg/L corresponding to Cadmium while the highest value is 0.425 mg/L belonging to Zinc. The treatment with chitosan was effective on the following metals: Cadmium 0.010 mg/L, Chromium 0.015 mg/L and Lead 0.159 mg/L. Zinc presented the same value.

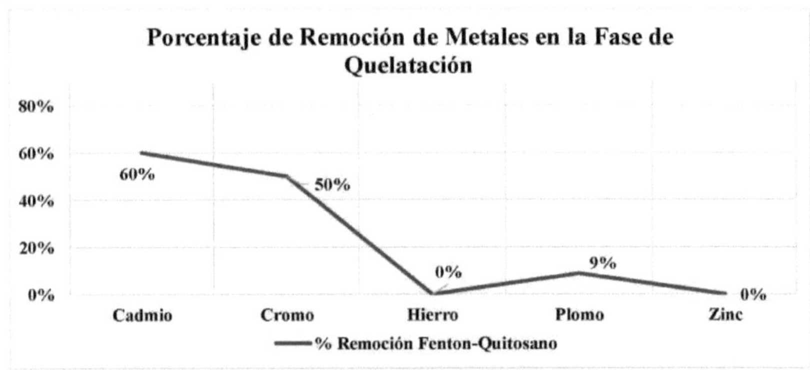


**Graph 11.-** Metals Removal Percentage

**Prepared by**: William Palacios, 2018.

The final percentages of heavy metal removal by applying Chitosan in the leachate treated with FENTON are: Cadmium 60%, Chromium 40% and Lead 9%. Iron and Zinc retained the same values.

# Chapter 3

# DISCUSSION

The leachate collected from the landfill pool has a pH of 8.5, indicating that it is an alkaline liquid. This collected leachate, being young, contains a high load of organic matter as well as heavy metals. The COD value measured was high at 9790 mg o2/L and exceeded the permissible limits by more than 100 %. By applying a treatment known as FENTON, the concentration of organic matter and substances susceptible to oxidation was lowered to 1015 mg o2/L, i.e. 91 % removal, although it is still outside the standard. The standard establishes that leachates to be discharged into fresh water bodies must have a maximum COD value of 200 mg o2/L. The production of chitosan, for its function as a chelating agent, was a determining factor in the execution of the project. For this project, 40.6 g of chitosan were produced from 200 g of shrimp shells, which represents 17 % of chitosan. The amount was sufficient for the chelation process in leachates.

The Chelation process decreased the concentration of three heavy metals presenting the following removal percentages; Cadmium 60%, Chromium 50% and Lead 9%. Iron and Zinc had 0% removal. Chitosan was effective in the removal of three of the five metals analyzed, of which Cadmium initially did not comply with the environmental standard, however, with the application of Chitosan, the value of this metal is within the maximum permissible limits. On the other hand, the treatment with the Fenton method was 91 % effective in the reduction of COD, which potentiated the effectiveness of Chitosan in this leachate.

Chitosan as an organic chelating agent was effective and economical, reducing the concentrations of heavy metals with a degree of efficiency of up to 60%, assisted by a pretreatment called FENTON. The metal that reached the maximum removal was chromium with 60% and the metal that had a minimum removal was lead with 9%.

The parameters established for the chelation and purity of the chitosan were determinant in the achievement of the objective of this study. The pH, the amount of chitosan and the agitation time used in this study can be modified to increase the effectiveness of the removal of heavy metals that present high concentrations. These results can be used as a reference for future trials since this project proved that chitosan did work as a heavy metal remover.

# CONCLUSIONS

The removal of heavy metals from the leachate of the Mejia landfill pool using chitosan was effective for Cadmium, Chromium and Lead. Zinc and iron were not removed due to their high concentration. In other words, Chitosan works on metals with low concentrations.

According to the effectiveness percentages of Chitosan in the removal of heavy metals, Cadmium is the metal that best fits the desired results because it presents a low concentration in the Fenton-treated leachate. This means that using Chitosan to remove Cadmium is reliable when its concentration is low.

The processes of deproteinization, demineralization, purification and deacetylation had notable losses in obtaining chitosan, obtaining only 17% pure chitosan. Likewise, the chitosan obtained from the shrimp exoskeleton had a deacetylation degree of 75%, and proved to be efficient in the elimination of Cadmium, Chromium and Lead ions at low concentrations.

On the other hand, the treatment of the raw leachate using the FENTON Advanced Oxidation method was able to eliminate 91 % of organic matter. It is worth mentioning that the leachate treated with FENTON was the one used for the chelation tests with chitosan.

The concentrations of heavy metals in the FENTON-treated leachate are:

Cadmium 0.025 mg/L, Chromium 0.03 mg/L, Iron 17.16 mg/L, Lead 0.174 mg/L, Zinc 0.425 mg/L. Chromium, lead and zinc comply with current environmental regulations.

Subsequently, Chelation with Chitosan was effective in the removal of the following metals. Cadmium 0.01 mg/L, Chromium 0.015 mg/L, Lead 0.159 mg/L. In this process four metals comply with the maximum permissible values of the current environmental regulations, these are: Cadmium, Chromium, Lead and Zinc. Although chitosan had no effect on zinc due to its high concentration, it was within the permissible limits. On the other hand, iron does not comply with the maximum limits of current environmental regulations and there was no reaction with chitosan due to the interference of ferrous sulfate in the treatment of the leachate with Fenton.

Chelation with chitosan as a chelating agent was effective on heavy metals with low concentrations

(Cadmium, Chromium and Lead), and was null on metals with high concentrations (Iron and Zinc). The percentages of effectiveness of Chitosan in the chelation of heavy metals in the leachate treated with FENTON are as follows: Cadmium 60%, Chromium 40% and Lead 9%. This shows that there was a higher effectiveness in Cadmium and a minimum in Lead. Chitosan removed 60% of the total heavy metals analyzed, especially those with low concentrations.

In mature leachates that present a lower concentration of heavy metals, it is recommended to use Chitosan directly in solution with acetic acid for greater effectiveness. In the case of young leachates, it is recommended to apply Chitosan powder after the FENTON Advanced Oxidation method to enhance its effectiveness in the removal of heavy metals. Additionally, to treat heavy metals with high concentrations in young leachates, it is recommended to modify the Chelation parameters, performing tests at different pH, agitation time and increasing the amount of Chitosan, using as a basis the results obtained in this project.

## ACKNOWLEDGMENTS


- The author would like to thank the administration of the "Romerillos" landfill in Cantón Mejía for providing the leachate samples.
- The author is grateful to SEK International University for the equipment, materials and reagents provided for the development of this project.

# REFERENCES CITED


- Altamirano, M. (2015). *"Removal of Pb $_{2+}$ by Adsorption on Chitosan"*. Faculty of Chemical Sciences. Universidad Veracruzana.

- Anrango, M. (2018). *"Alternative Methods for the Treatment of Leachate from the Mejía Canton Landfill, Pichincha, Ecuador"*. Faculty of Natural and Environmental Sciences. Sek International University.

- APHA AWWA WEF. (2005). *"StandardMethods for the Examination of Water and Wastewater"*. 21st edition. Metals part 3000. 3010A p. 3-1

- Arevalo, C. (2018). *"Evaluation of the hydraulic behavior of leachate from the north phase I of the Cuenca landfill"*.

    - ATSDR, (2004). *"Public Health Summary, Cobalt."*

- Balleño, A., Ríos, N., Aranda, F., Morales, J., Mendizábal, E & Katime, K. (2016). *"Alginate-Chitosan and Alginate-Chitosan Sulfate Hydrogels for the Removal of Copper Ions"*. Centro Universitario de Ciencias Exactas e Ingenierías. University of Guadalajara. Guadalajara. Guadalajara. Mexico.

- Blanco, J. (2009). *"Degradation of a real textile effluent by Fenton and Photo-Fenton processes"*. Polytechnic University of Catalonia.

- Cahuasqui, S. (2011). *"Determination of heavy metals (lead, cadmium and nickel) in coriander (Coriandrum sativum L) in Aloag, Canton Mejia, Province of Pichincha by flame atomic absorption spectrophotometry"*. Pontificia Universidad Católica del Ecuador Faculty of Exact and Natural Sciences School of Chemical Sciences.

- Cañizares, R. (200). *"Biosorption of Heavy Metals using Microbial Biomass"*.

    - Coral, K. (2011). *"Solid Waste Treatment"*. UISEK.

- Davila, G & Bonilla, P. (2011). *"Optimization of the Lead Adsorption Process with Chitosan"*. Faculty of Chemical Sciences, Universidad Central del Ecuador.

- Diaz, F. (2012). *"Removal of Hexavalent Chromium from Contaminated Waters using Chitosan obtained from Shrimp Exoskeleton"*.

- Eróstegui, C. (2009). *"Contamination by Heavy Metals"*. Volume 12. Neurophysiology Editor.

- Escobar, D., Ossa, C., Quintana, M., & Ospina, W. (2013). *"Optimization of a protocol for the extraction of chitin and chitosan from crustacean shells."* Scientia Et Technica, vol. 18, pp. 260-266 Universidad Tecnológica de Pereira.

- Garcia, J. (2014). *"Fenton and photo-fenton processes for the treatment of microbiological laboratory wastewater employing $Fe_2O_3$ "*. Pontificia Universidad Javeriana. Bogotá.

- Giraldo, E. (2001). Landfill leachate treatment: recent advances. *Revista de ingeniería*, (14), 44-55.

- Hernández, H. (2009). *"Obtaining and characterization of chitosan from shrimp exoskeletons"*. Faculty of Chemical Engineering, Benemérita Universidad Autónoma de Puebla Avenida San Claudio.

- Jiménez, D. (2000). *"Evaluation of the parameters of a continuous type Anaerobic Biodigester"*. Faculty of Electrical Mechanical Engineering.

- Knoch, J., & Stegmann, R. (1993). Leachate treatment. In *Leachate treatment*. CAPRE/ANDESAPA.

- Plan de Desarrollo y Ordenamiento Territorial del Cantón Mejía. (n.d.). Pichincha, Ecuador: Mejía Municipal Government.

- Pellon, A., Matilde, L., & Espinosa, M. d. (2009). Technology for the Treatment of Leachate from Municipal Solid Waste Landfills. *Tecnología Química* , 113-121.

- Pontificia Universidad Católica de Valparaiso (s/f). Anexo Planta de Tratamiento de lixiviados. Chile: Ilustre Municipalidad de Castro.

- Lárez, C. (2008). *"Some potentialities of chitin and chitosan for agriculture-related uses in Latin America.*" Polymer Laboratory, Department of Chemistry, Faculty of Sciences, Universidad de Los Andes. Mérida 5101, Venezuela. E-mail: clarez@ula.ve

- Lozada, P. T., Rodríguez, J. A., Barba, L. E., Morán, A., & Narváez, J. (2005). Anaerobic leachate treatment in UASB reactors. *Engineering and Development*, (18).

- Martínez, O. (2009). *Improvements in the treatment of MSW landfill leachate by advanced oxidation processes.* University of Cantabria.

- Nájera, H., Castañón, J., Figueroa, J., & Rojas-Valencia, M. (2009). Characterization and physicochemical treatment of mature leachate produced at the Tuxtla Gutierrez disposal site. *Chiapas, Mexico*, 9.

  - Nordberg, G. (2007). "*Metals: Chemical Properties and Toxicity.*

- Rosales, J. (2008). "*Treatment of leachate produced in the Inga landfill by coagulation, flocculation and anaerobic treatment processes at laboratory scale*". SEK International University.

- Rubio, A. (2014). "*Application of the Fenton process in the treatment of petrochemical wastewater*". University of Antioquia, Medellin, Colombia. ** Department of Mechanical Engineering. University of Antioquia, Medellín, Colombia.

- Salas, G. (2010). "*Advanced Oxidation Treatment (Fenton reaction) of wastewater from the textile industry.*" vol. 13 n.0 1, 2010. pp. 30-38.

- Velasco, J. (2002). "*Tecnologías de remediación para Suelos*". National Institute of Ecology.

- Villalobos, W. (2011). ^*Elumination of Heavy Metals from Wastewater with Chitosan Membranes"*, School of Chemistry. Technological Institute of Costa Rica.

- Zulay, M. (2011). "*Chitin and chitosan friendly polymers. A review of their applications*". Revista Tecnocientífica URU Universidad Rafael Urdaneta Facultad de Ingeniería.

- Zaldumbide, L. (2012). *"Physical Characterization of Urban Solid Waste, Chemical Characterization of Leachate and Treatment Proposal for Leachate from the Mejía Canton Landfill"*. SEK International University, Faculty of Environmental Sciences.

# I want morebooks!

Buy your books fast and straightforward online - at one of world's fastest growing online book stores! Environmentally sound due to Print-on-Demand technologies.

Buy your books online at
**www.morebooks.shop**

Kaufen Sie Ihre Bücher schnell und unkompliziert online – auf einer der am schnellsten wachsenden Buchhandelsplattformen weltweit! Dank Print-On-Demand umwelt- und ressourcenschonend produzi ert.

Bücher schneller online kaufen
**www.morebooks.shop**